Attrassi Khaled

Controlling Citrus Fruit Rot

Effect of two essential oils

ScienciaScripts

Imprint
Any brand names and product names mentioned in this book are subject to trademark, brand or patent protection and are trademarks or registered trademarks of their respective holders. The use of brand names, product names, common names, trade names, product descriptions etc. even without a particular marking in this work is in no way to be construed to mean that such names may be regarded as unrestricted in respect of trademark and brand protection legislation and could thus be used by anyone.

Cover image: www.ingimage.com

This book is a translation from the original published under ISBN 978-620-6-69083-2.

Publisher:
Sciencia Scripts
is a trademark of
Dodo Books Indian Ocean Ltd. and OmniScriptum S.R.L publishing group

120 High Road, East Finchley, London, N2 9ED, United Kingdom
Str. Armeneasca 28/1, office 1, Chisinau MD-2012, Republic of Moldova, Europe
Printed at: see last page
ISBN: 978-620-6-49841-4

Attrassi Khaled

Controlling Citrus Fruit Rot

CONTROL OF CITRUS FRUIT ROT EFFECT OF TWO ESSENTIAL OILS

Dr. Khaled ATTRASSI

Table of contents

GENERAL INTRODUCTION

GENERAL INFORMATION ON CITRUS

The word citrus comes from Latin, and has been used since antiquity to designate trees with acid fruits (Guillaume, 2011).

For 3,000 years, citrus fruits have been considered ornamental plants and grown in the orangeries of castles, where the beauty of the foliage is added to that of the flowers and fruit; but today they are grown all over the world, for their food use given the high nutritional, gustatory and organoleptic quality of their fruits and their beneficial effects on human health (Benaissat, 2015).

Citrus originated in southern Asia, particularly in the lower Himalayas, where hot, humid climates dominate. Over time, man's many conquests and travels have brought these species to Europe, Africa and then America, where Mediterranean and subtropical climates have proved to be more favorable to the cultivation of these plants. (Guillaume, 2011).

HISTORY OF CITRUS GROWING

Citrus fruits belong to the Rutaceae family. This family comprises three main genera: Citrus sp: *Citrus* sp., *Poncirus* sp. and *Fortunelles* sp.; Each genus has its own origin, the *Citrus* genus has a southern origin, between India and Indonesia, while the other two genera, *Poncirus* and *Fortunella,* originate in the northern areas of eastern China (Luro, 2015).

It's the *Citrus* genus that contains the vast majority of species, its diversity is at the origin of four taxonomic units, *Citrus maxima* (the grapefruit trees), *C. reticulata* (the mandarin trees), *C. medica* (the citron trees) and the papedas (Combava) (*C. hystrix*\ from which the majority of cultivated species have emerged. The first three species, are considered according to Luro (2015) to be ancestral species, which appeared and evolved separately in three distinct geographical areas, respectively, the Malaysian archipelago, southern China and northeastern India).

The appearance of the other species was the result of a series of sexual crosses between ancestral species produced in mixed stand areas, resulting in hybrid, interspecific forms such as: orange *(C. sinensis)* and sour orange (*C. aurantium),* produced from crosses between grapefruit and mandarin trees, the lemon tree *(C. limon)*, a hybrid of citron and sour orange and finally the lime tree *(C. aurantifolia)* produced from a cross between a papeda *(C. micrantha)* and a citron tree (Luro, 2015) (Figure 1).

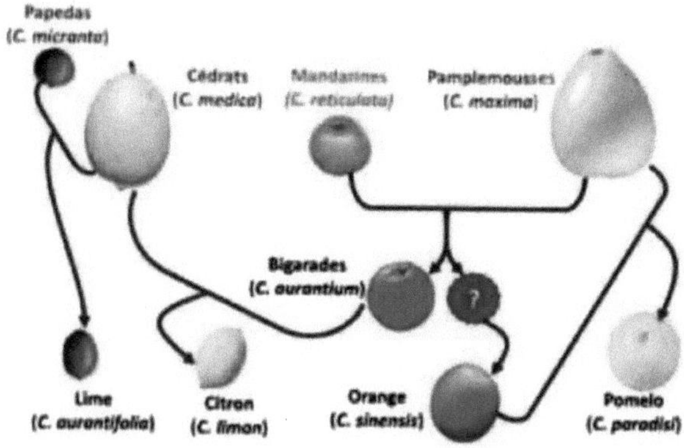

Figure 1: Phylogenetic tree of species in the Citrus genus (Luro, 2015).

The first major migration of citrus fruit took place in the first millennium BC, heading west to Mesopotamia, between the 8th and 4th centuries BC. The first migrant of these fruits was the citron tree, introduced by Alexander the Great to the Sumerians and Egyptians. From there, and at the beginning of the Christian era, they reached the shores of the Mediterranean, thanks to the Israelites who used them as an offering during the Feast of Tabernacles. Thanks to the Moors, Genoese and Portuguese, bigarad and grapefruit were introduced to the Mediterranean between the 10th and 12th centuries. Then came the turn of the lemon tree, which was spread by the Arabs around the Mediterranean from the 10th century onwards. The rise of maritime trade in the 15th century enabled citrus to spread around the world (Luro, 2015).

Sweet oranges arrived in Europe in the early 15th century, thanks to the Portuguese, following the discovery of the route to India via southern Africa. The mandarin tree was introduced to the West in the 18th century, through the gateway to Great Britain (Luro, 2015).

Today, citrus fruits are grown in all the world's mild-climate regions: pomelos from Florida, oranges from South Africa, large grapefruits from China and mandarins from Corsica. However, we can expect to see fruits from other countries, as new citrus fruits are still being discovered at the start of the 21st century (Virbel-Alonso, 2011).

The genera *Poncirus* and *Fortunella* produce non-edible fruit; it is the *Citrus* genus that contains the vast majority of cultivated citrus species. There are two different classifications for this genus, that of Tanaka (1961), which includes 156 species, and that of Swingle and Reece (1967), which contains only 16 species, of which only eight are cultivated (Luro et *al.,* 2001). This great difference is explained by the great potential for interspecific hybridization, not to mention the polyembryony that enables these hybrid structures to be fixed (De Rocca et *al.,* 1992).

Table 1: Citrus classification (Praloran, 1971; Chase, 2003; Groppo et al. *al.,* 2008).

	Classical classification	**Phylogenetic classification**
Branch	Angiosperms	Angiosperms
class	Eudicotes	Eudicotes
Subclass	Archichlomydeae	Rosideae Eurosideae
Order **Family**	Germinales (Rutales) Rutaceae	Sapindales Rutaceae
Sub-family	Aurantioideae	Aurantioideae
Tribes	Citreae	Citreae
Sub-tribe	Citrinae	
Genres	*Poncirus, Fortunella and Citrus.*	*Poncirus, Microcitrus, Citrus, Atalancia and Naring*

Citrus is a group of small trees, often thorny, with alternate, simple, unstipulated leaves, yellow translucent glands and terminal inflorescences. The flowers are hermaphrodite, actinomorphic, with 4 or 5 free or fused sepals, 4 or 5 free petals; the androecium contains 8 to 10 stamens inserted on a nectariferous disk; the gynoecium has a supere ovary with 4 to 5 dehiscent chambers. (Quezel et Santa, 1963; Fennane et *al.,* 1999). The fruit is generally round or ovoid (Guillaume, 2011).

SPECIES AND VARIETIES

Among the best-known citrus species in the world of agriculture, there are 9: orange, mandarin, lemon, grapefruit, clementine, lime, sour orange, citron and pomelo.

Table 2: Most commonly grown citrus species and their descriptions.

Species	Description
 (Mekkioui, 2015)	The orange tree is undoubtedly the most widely grown and best-known citrus fruit in the world. It is a medium-sized, spherical fruit (Guillaume, 2011).
 (site 2)	The bitter orange tree is an inedible fruit, due to its acidic and very bitter taste, which makes it unfit for consumption (Guillaume, 2011).

 (Mekkioui, 2015)	Grapefruit trees can reach and even exceed 10 m in height. They are characterized by a coarse pulp, a marked placental void and mono-embryonated seeds (Brebion et *al.*, 1999).
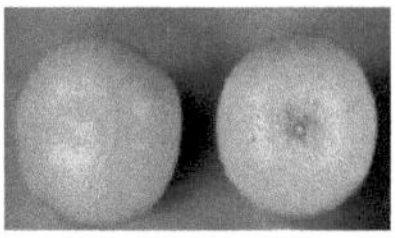 (Mekkioui, 2015)	Grapefruit is a hybridization or mutation of grapefruit, as it differs from the latter in having much smaller fruits, a thinner rind, juicy pulp and polyembryonated seeds (Brebion et *al.*, 1999).
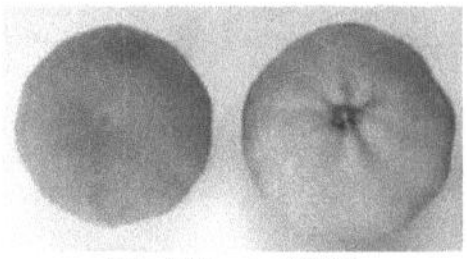 (Mekkioui, 2015)	Mandarins are particularly fragrant and sweet, globose fruits often flattened at both poles and containing numerous seeds (Brebion et *al.*, 1999).
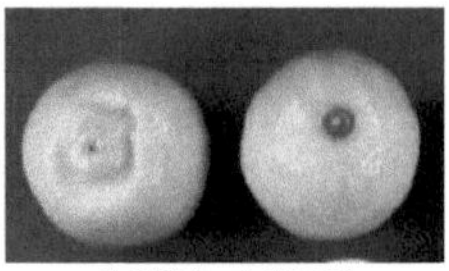 (Mekkioui, 2015)	Lemons are ovoid, yellow fruits with a fine, juicy, acidic pulp. Ripe, yellow fruits can remain on the tree for several months without losing their quality (Brebion et *al.*, 1999; Guillaume, 2011).

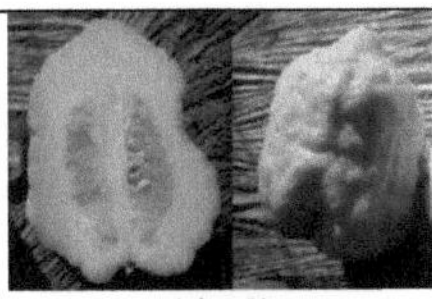 (site 3)	The citron tree is distinguished from other species by the considerable thickness of its fruit rind. These large fruits, weighing up to 4 kg, are sensitive to cold and heat (Brebion et *al.,* 1999).
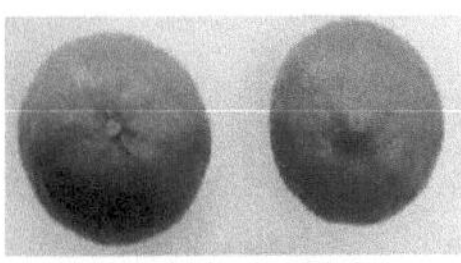 (Mekkioui, 2015)	Lime trees are thorny shrubs with small, elliptical, pale-green leaves. Grown in tropical climates, they are distinguished from other species by their juicy, highly acidic, greenish pulp (Brebion et *al.*, 1999).
 (Mekkioui, 2015)	The history of the Clementine tree goes back to 1902, in Algeria, with a small intra-specific hybridization between a mandarin and a bigaradier that resulted in the production of a new species, named after their inventor, the Clementine brothers (Brebion et *al.*, 1999).

DIFFERENT FUNGAL DISEASES OF CITRUS FRUIT IN STORAGE

Citrus is attacked by a number of pathogens from flowering to the harvest stage and subsequently by post-harvest pathogens, most notably fungi, which affect crop production and significantly deteriorate fruit quality (Embaby et *al.,* 2013).

Several types of injury can affect post-harvest fruit quality. Damaged tissue becomes susceptible to infection by pathogenic microorganisms (Fischer et *al.,* 2009).

Infection and contamination occur at various stages in the field and after harvest during marketing. Most post-harvest pathogens *(Penicillium, Alternaria, Diplodia* and *Phomopsis)* are able to invade wounds when host defenses are weak. Microorganisms are highly selective in host tissues, depending on pH values (acidic, basic or neutral). Citrus fruits have a pH below 4, so most fungi attack these fruits (Olsen et *al.,* 2000; Milind, 2008).

There are several so-called typical post-harvest diseases, such as green mold, blue mold, *Alternaria* black rot and stem rot, which also reduce fruit quality (Fischer et *al.,* 2009; Embaby et *al.,* 2013).

Green mold (*P. digitatum)* and blue mold (*P. italicum)* are important post-harvest pathogens in citrus. However, both green and blue molds occur in all citrus growing zones and are often the predominant type of rot. (Snowdon, 1990; Naqvi, 2004). These fungi are long-lived and fairly durable, even under adverse conditions. *Penicillium* growth typically occurs following wound infections in produce (Nunes et *al.,* 2010). *Penicillium,* in general, are post-harvest pathogens and a major problem for Mediterranean orchards and the fruit industry (Munoz et *al.,* 2011). Green and blue moulds, respectively, are the main cause of post-harvest losses in oranges. Under Mediterranean climatic conditions, the two together are responsible for 80% of total post-harvest citrus decline (Nunes et *al.,* 2010; Munoz *et al.,* 2011).

Citrus black rot, caused by *Alternaria,* can occur in the field before harvest. The disease is a common problem in Mediterranean climates due to cool, wet winters and hot, dry summers (Brown and Eckert, 2000). In one of the earliest descriptions of *Alternaria* rot of citrus, the causal agent was identified as *Alternaria citri* (Embaby et *al.,* 2013). *Alternaria* rot develops most rapidly in cold storage, and the fungus grows internally along the central core of the fruit. *Alternaria* rot has been considered a serious problem in lemons and oranges for long-term storage (Embaby et *al.,* 2013).

Citrus rots caused by *Phomopsis citri* are the main postharvest diseases in the citrus-growing area that receive significant rainfall during the fruit development period (Embaby et *al.,* 2013).

FUNGAL SPOILAGE OF CITRUS FRUIT

The wide diversity of citrus varieties and climates in production and consumption areas attract an equal diversity of pathogens both in orchards and at conservation stations (Benaissat, 2015).

Post-harvest diseases can have a significant impact on maintaining the quality of fresh citrus fruit. Decay levels can often reach 20-40% if these fruits are not treated co rrectly with fungicides and/or preserved by cold storage (Brown and Eckert, 1988).

The incidence of spoilage is highly variable and depends on a number of factors, including mould type, fruit variety, environmental storage conditions, harvesting procedures, packaging conditions, fungicide efficacy and application method (Brown and Eckert, 1988).

Molds responsible for postharvest disease in citrus may be inactive for a long period after infection, before causing disease symptoms, or symptoms may develop soon after infection (Brown and Eckert, 1988). Among the fungi responsible for postharvest disease in citrus, we can distinguish three genera: *Penicillium sp., Aspergillus sp.* and *Acremonium sp.*

PENICILLIUM SP.

This genus comprises filamentous, imperfect fungi belonging to the Deuteromycetes division. A few perfect (sexual) forms are known and belong to the Ascomycetes division *(Carpenteles, Eupenicillium, Talaromyces).* More specifically, *Penicillium* belong to the class *Eurotiomycetes,* order *Eurotiales,* family *Trichocomaceae* and genus *Penicillium* (Webster and Weber, 2007). The genus currently comprises around 354 species (Visagie et *al.,* 2014).

Penicillium is a diverse genus occurring worldwide and its species can be responsible for numerous degradations; they are also capable of producing a wide range of mycotoxins (Visagie et *al.,* 2014).

Penicillium species are defined essentially by the characteristics of the thallus, penicils and spores (Tabuc, 2007). The characters of conidiophores and cleistothecia are of great taxonomic importance (Figure 2). Other important microscopic characters include conidiophore wall texture, stipe and conidial ornamentation, dimensions, and sometimes colors of all conidiophore elements are thus involved in classification (Visagie et *al.,* 2014).

However, classifying and identifying *Penicillium* by phenotypic tools alone is difficult to achieve since they are often very similar and their phenotypic plasticity is great (Labonté, 2012). Currently, on the molecular side, there is no comprehensive and verified reference database for the genus (Visagie et *al.,* 2014).

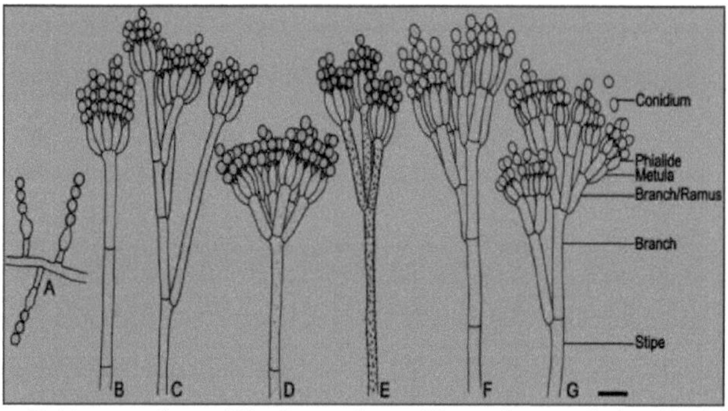

Figure 2: Patterns of conidiophore branching observed in *Penicillium.* A. Conidiophores with solitary phialides. B. Monoverticulate. C. Divaricate. D, E. Biverticillé. F. Terverticillé. G. Quaterverticillate. Scale bar = 10 µm (Visagie et *al.,* 2014).

ASPERGILLUS SP.

The genus *Aspergillus* belongs to the class Ascomycetes, with a hyaline or colored thallus, comprising septate mycelium bearing several erect conidiophores, which terminate in vesicles (Figure 3) (Raper and Fennell, 1965)-Phialides are borne by metules attached to the surface of the vesicles (Dufresne Guy St-Germain, 2014).

The genus *Aspergillus* comprises 339 species (Samson et al. 2011), these species are classified in ten different teleomorph genera, showing that Aspergillus form a very diverse group of fungi (Krijgsheld *et al.,* 2013).

Aspergillus are most often found in warm climates (Castegnaro and Pfohl-Leszkowicz, 2002).

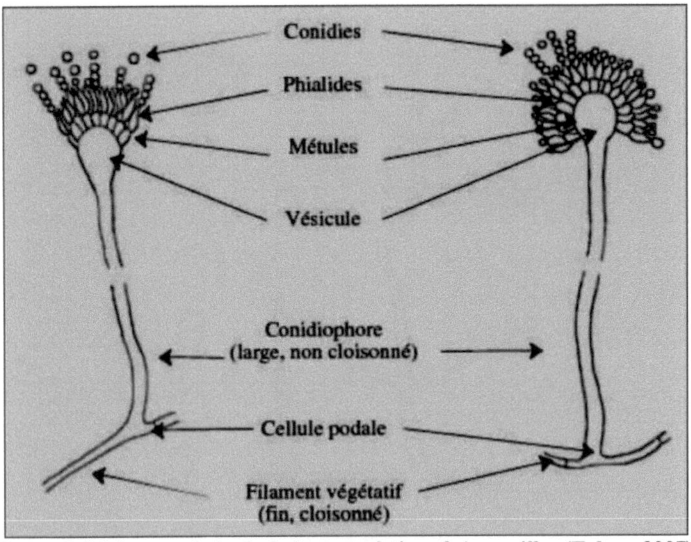

Figure 3: Main morphological characteristics of *Aspergillus* (Tabuc, 2007).

ACREMONIUM SP.

Acremonium is a genus of anamorph molds; it has a filamentous, simply structured thallus (Summerbell et *al.,* 2011); it currently comprises nearly 200 different species (Hordé, 2015). Most of them are saprophytic and are isolated mainly from dead plant material and soil. They are found almost everywhere in nature (Giraldo et *al.,* 2012); some *Acremonium* species live entirely within the vegetative and reproductive organs of Poaceae (Raynal, 1991).

Some species can cause opportunistic infections in humans and animals. Cultures of *Acremonium* species are generally slow-growing, their characteristic morphology consisting of creeping, septate hyphae giving rise to slender, narrow, tapering phialides. Although conidiophores, differentiated with or without whorled branching, can be observed in some species (Figure 4). Conidia are generally small, unicellular and aggregated in mucoid heads or unbound chains. Sometimes heads and chains are present in a single culture. They may be hyaline or melanized, but hyphae are generally hyaline. Characteristic chlamydospores and sclerotia are produced in some species (Summerbell et *al.,* 2011; Giraldo et *al.,* 2012).

Figure 4: ***Acremonium*** **conidiophore. A and B: conidiophores; C: conidia attached to a phialid; D: conidia.**

HOW TO CONTROL THE FUNGI RESPONSIBLE FOR CITRUS FRUIT DETERIORATION

The most effective method for controlling citrus molds in general is chemical control, which means using synthetic products with a harmful effect on fungi; this method is the most widely used in citrus preservation. However, there are other biological methods that are no less important than chemical control, such as the use of biopesticides, particularly essential oils, which have proven antifungal properties on certain fungi.

ESSENTIAL OILS

Essential oils (EO) are a set of volatile, fragrant compounds extracted from the tissues of certain plants; they are more or less complex compounds, containing a mixture of molecules such as terpenes, phenols, methyl ethers, oxides, esters, ketones, ... (bouguerra, 2012); they are formed in some plants as products of secondary metabolism (Alilou, 2012). EOs are mainly involved by plants in defense strategies against pathogenic organisms (Csesk and Kaufman, 1999).

EOs are present in various plant organs, varying from one species to another depending on the plant's production zone (Lakhdar, 2015). They are generally accumulated in isolated cells and are characterized by their often yellow hue and thick walls (Benayad, 2013). EOs are produced in small quantities in plants (0.01 to 5% of their weight) (Benbouali, 2006).

The role of EOs in general remains obscure, yet they can have an effect on aromatic plants themselves, as well as on consumers in the food chain. They can act as a pollination guide, attracting pollinating insects with their strong odors, or as a pest repellent (Benbouali, 2006).

ESSENTIAL OIL EXTRACTION PROCESSES

There are several methods for extracting EOs, each with its own principle, while hydrodistillation and steam distillation remain the most widely used techniques on an industrial scale (Wang et *al.,* 2008). Other techniques include cold expression, volatile organic solvent extraction, superheated steam extraction and supercritical CO_2 extraction (Bruneton, 1999).

Hydrodistillation is the oldest and most cost-effective technique for extracting EOs. Several studies have used this technique to extract EOs from thyme (Akrout, 2004; El Ajjouri et *al.,* 2008; El-Akhal et *al.,* 2014) and spearmint (Lahrech 2010; Hmiri et *al.,* 2011; Foganholi et *al.,* 2015); this method has the advantage of not requiring a lot of equipment, and is not expensive (El Ajjouri, 2013).

The extraction of thyme EOs has been the subject of several experimental studies, one of which found that extraction using supercritical CO_2 at low temperature and high pressure gave good results compared with steam stripping (Maldao-Martin et *al.,* 2000).

AZOXYSTROBIN

Azoxystrobin is a broad-spectrum systemic fungicide belonging to the strobilurin family, developed by Zeneca UK Ltd. in 1992 (syngenta, 2008).

Strobilurins have a totally different mode of action to triazoles, with powerful plant distribution properties and a broad spectrum of action, as they act at cytochrome b complex sites. Strobilurins open up new prospects in the fungicide market. They act on various pathogens belonging to the Ascomycetes, Basidiomycetes, Deuteromycetes and Oomycetes (Ypema, 1999; Wong, 2001).

Azoxystrobin has been registered mainly for rice, wheat, beans and grapes, etc. in around 50 countries (Food Safety Commission, 2006).

In Morocco, this molecule is registered to protect vines, tomatoes, peppers and Cucurbitaceae against powdery mildew, downy mildew and alternaria. On cereals, it is used against rusts, septorioses and soon against helminthosporioses and fusarioses of ears (Bahous et *al.,* 2005).

DESCRIPTION OF THE ACTIVE INGREDIENT

The table below briefly describes azoxystrobin.

Table 3: Brief description of azoxystrobin (Food Safety Commission, 2006).

Common name	Azoxystrobin
Chemical name	Methyl (*E*)-2-{2-[6-(2-cyanophenoxy) pyrimidin-4-yloxy]phenyl}-3-methoxyacrylate.
Molecular formula	C22H17N3O5

Chemical structure	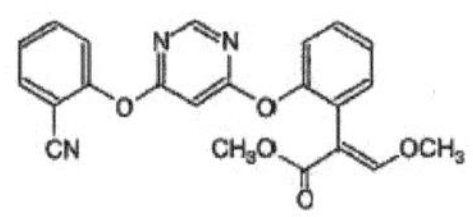

PHYSICAL AND CHEMICAL PROPERTIES OF AZOXYSTROBIN

Azoxystrobin has the property of being transported by the xylem, as it diffuses homogeneously around impact points; it can also be put back into solution by being redistributed by rain or dew (syngenta, 2008).

Azoxystrobin has a spray drying time of 1 to 2 hours, giving it excellent resistance to leaching, and its efficacy is independent of temperature, so it can remain active for two weeks (syngenta, 2008).

Other physico-chemical properties of azoxystrobin are summarized in Table 4.

Table 5: Physicochemical properties of azoxystrobin (New Strategies and Regulatory Affairs Division, 2007).

Properties	Values	Comments
Molecular weight [g/mol]	403,4	
Physical condition and Color	White powder	
Odor	No characteristic odor	

Density at 25 °C	1.25 g/ml	
Vapour pressure at 20 °C	1.1×10^{-li} kPa	Relatively non-volatile under natural conditions
Henry's law constant at 20°C	7.310⁻ " ATMm3/mol $1/H = 3,3 \times 10^{11}$	Low volatilization potential from water bodies and wet surfaces
Solubility in water at 20 °C	6.0 mg/L	Low water solubility
Melting point	116 °C	

AZOXYSTROBIN MODE OF ACTION

Azoxystrobin penetrates host tissues to halt fungal development and block disease progression (Bahous et *al.,* 2006).

Azoxystrobin is mainly involved in spore germination and sporulation, and to a lesser extent in mycelial growth, the most energy-intensive stages of fungal development (syngenta, 2008).

Their main role is to inhibit the electron transport system by binding to the Qo site of the mitochondrial cytochrome b complex, thus disrupting respiration, and rather the energy cycle within the fungus, and thus rapidly halting its development (Bahous et *al.,* 2006).

Azoxystrobin's powerful anti-germinative action makes it essentially preventive. Its anti-sporulation action prevents further contamination (syngenta, 2008).

PHYSIOLOGICAL STUDY OF CITRUS FRUIT ROT DURING STORAGE IN MOROCCO

PHYSIOLOGICAL STUDY OF THE ROT OF CITRUS FRUIT IN CONSERVATION IN MOROCCO

Summary

Citrus fruit is harvested at the start of ripening to maximize shelf life and marketability. However, prolonged storage of apples exposes them to a variety of ailments, which can be physiological in origin, often linked to cold, or parasitic in origin, mainly fungal diseases.

Not all fruit entering the storage facility becomes marketable, resulting in financial losses in addition to transport and storage costs. It is difficult to estimate losses due to individual pathogens, as wounds are colonized by a complex of fungi.

A physiological study of the effect of two culture media on the development of four fungal species *(Penicillium italicum, Penicillium digitatum, Aspergillus niger* and *Acremonium sp.)* isolated from rotted citrus fruit during storage showed that MEA and PDA media are favorable to the development of these fungi. On PDA medium, mycelial growth is greatest at 20°C.

Keywords: Citrus, Decay, Identification, Physiological study.

INTRODUCTION

In Morocco, as in other countries, citrus fruits are classified as essential agricultural products, constituting national resources that can provide income and employment opportunities in rural and peri-urban areas. What's more, as permanent crops, they help create more stable agroecosystems and protect the environment (FAO, 2003).

According to 2013 statistics from the Food and Agriculture Organization of the United Nations (FAO) (figure a), the world's leading citrus-producing country is China, with a total of 33,104,744 tonnes, followed by Brazil with 19,734,725 tonnes, then the United States of America with 10,133,246 tonnes, India is ranked 4ème with 10,090,000 tonnes of production; according to the same statistics, Morocco comes in 15ème with a total of 1,467,925 tonnes.

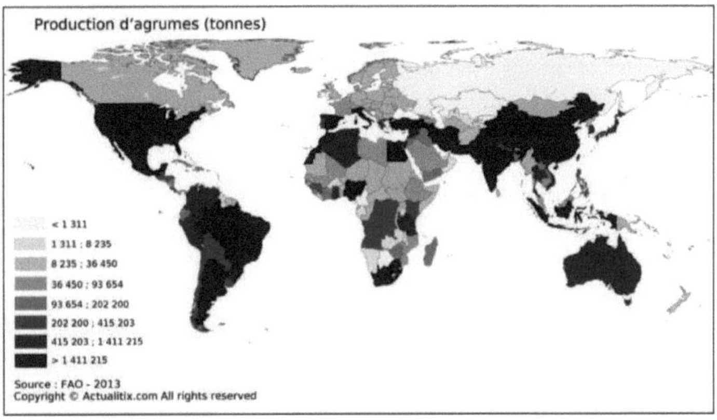

Figure a: World citrus production statistics by tons in 2013.
The color deepens as production increases: dark blue means production in excess of 1,411,215 hectares, and so on down to production of less than 1,311 hectares in white.

In Morocco, on a citrus-growing area of up to 118,000 ha, including 92,000 ha of productive area (Figure b), citrus production for the 2013-2014 crop year was 2.2 million tonnes, representing an average yield of 24 tonnes/ha (Direction de la Stratégie et des Statistiques, 2014).

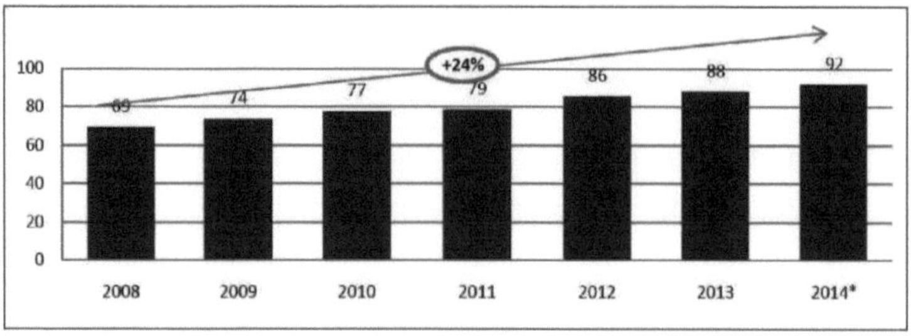

Figure b: Evolution of productive citrus area (Kha) between 2008-2014 (Ministère de l'Agriculture et de la Pêche Maritime, 2014).

Moroccan citrus production has exploded since the launch of the "Green Morocco" plan, which saw production pass the two-million-ton mark in 2016, up from 1.3 million in 2008; but despite this progress, the Green Plan has not influenced the export field, which has remained at the same level as in the early 2000s, with nearly 500,000 tons exported each year (Dilami, 2016).

The pace of development of the citrus sector in Morocco has increased at an astonishing rate over the last few decades, thanks to the efforts made by farmers and large-scale producers to improve product quality in order to enhance its place on the international market, with a view to presenting a competitive product. Between 1962 and 2014, citrus production made great strides, thanks to favorable soil and climate conditions and new incentives.

In 1962, total production was 504,700 tonnes (Chapot et *al,* 1964), this production has risen to 2.2 million tonnes in 2014 (Ministère de l'Agriculture et de la Pêche Maritime, 2014).

The area under citrus production in Morocco is distributed randomly across the Kingdom's different regions, with 37% located in the Souss-Massa region, followed by the Gharb-Loukkos region with 19%, then the Oriental with 17%, Tadla with 15% and Haouz with 6%. The remaining 5% of the area is located in the regions of Taounate, Khémissat, Fès and Meknès (Ministère de l'Agriculture et de la Pèche Maritime, 2012).

The main varieties currently grown in the national citrus orchard are : Clementine (26%), Maroc Late (24%), Navel (19%), the remaining 31% are represented by Nour,

Nules, Afourer, half-season varieties and others (Ministry of Agriculture and Maritime Fisheries, 2012).

In this work, we studied the biology and physiology of four fungal species, isolated from citrus fruit lesions preserved in cold storage in Kenitra.

MATERIALS AND METHODS

Isolation and identification of fungal species

Rotten oranges from various markets in Kenitra were disinfected with 75° alcohol. Fragments showing lesions were then cut out and placed in the center of Petri dishes containing PDA medium supplemented with chloramphenicol.

Once the fungal colonies are well developed, a fragment of each colony is subcultured several times in new PDA dishes to obtain pure cultures. Microscope slides are prepared from the pure cultures and observed under a light microscope. The slides are mounted in a drop of distilled water. Isolated fungal species are identified on the basis of their morphological and microscopic characteristics (Samson et *al.,* 2004). Images were taken using an optical microscope (Leica DM750 with built-in camera).

Physiological studies

To determine some of the physiological parameters of the strains tested, we studied the influence of temperature and the composition of the culture medium on their growth.

a-Growing environments

Two culture media were used to grow molds isolated from citrus fruit: PDA and MEA.

PDA (Potato-Dextrose-Agar) medium consists of 200 g potato, 15 g agar and 20 g sucrose, and is mixed in an Erlenmeyer flask with 1000 ml distilled water.

The MEA medium consists of 15 g malt extract, 15 g agar and 1000 ml distilled water (Pitt, 1988; Botton et *al.,* 1990; Davet and Roxel, 1997).

After sterilization in an autoclave at 120°C for 20 min, both media are poured into Petri dishes and inoculated with fungal thallus fragments after solidification, followed by incubation in an oven at 25°C until fungal colonies appear.

If necessary, culture media are supplemented with 100 mg/l chloramphenicol as an antibiotic to inhibit bacterial proliferation.

b-Effect of culture medium composition on mold growth

Two culture media are being tested to assess fungal mycelial growth. These are PDA and MEA media.

For each mold, a 5 cm diameter disc of fungal culture is placed in the center of a Petri dish containing the culture medium to be tested. The cultures are then incubated in an oven at 25°C and fungal colony diameter measurements are taken daily for 14 days, each test being repeated 3 times.

c-Effect of temperature on mold growth

Mycelial disks from ten-day-old cultures are placed in Petri dishes containing PDA medium and incubated at different temperatures (4, 10, 15, 20, 25 and 30°C). Diameter measurements are taken every 24 hours to test the influence of temperature on the growth of each mold. Each test is repeated three times.

STATISTICAL ANALYSIS

Statistical processing of the data was carried out using XLSTAT software, with the "Duncan" test at the 5% threshold.

RESULTS AND DISCUSSION

We were able to isolate and identify several fungal agents responsible for citrus rots from citrus fruit that had rotted during storage. Ten fungal species were identified from the lesions of four rotten oranges: *Penicillium italicum, Penicillium digitatum, Aspergillus niger* and *Acremonium sp*. The identity of all the fungal species isolated will be confirmed by phyllogenetic studies.

Effect of culture medium

The graphs below (Figure 1 and 2) present the results of the effect of culture media on the growth of four species of fungi. They show that PDA and MEA media are the most favorable for mycelial growth of the majority of molds; these media allow good growth of almost all *P.italicum and P.digitatum;* however, *Acremonium sp* and *A.niger* have average growth on both culture media.

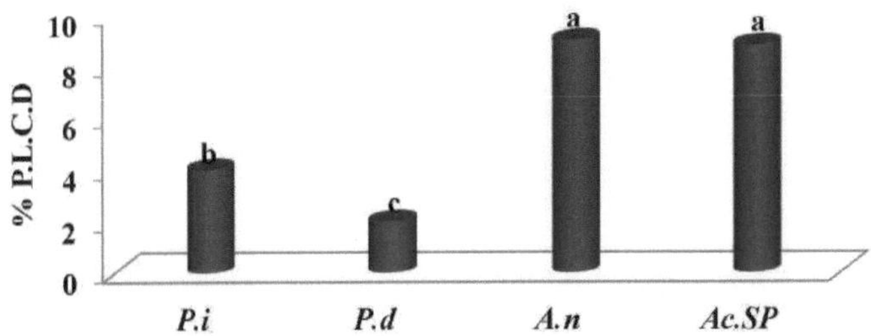

Figure 1: Influence of MEA culture media on the mycelial growth of four fungal species.

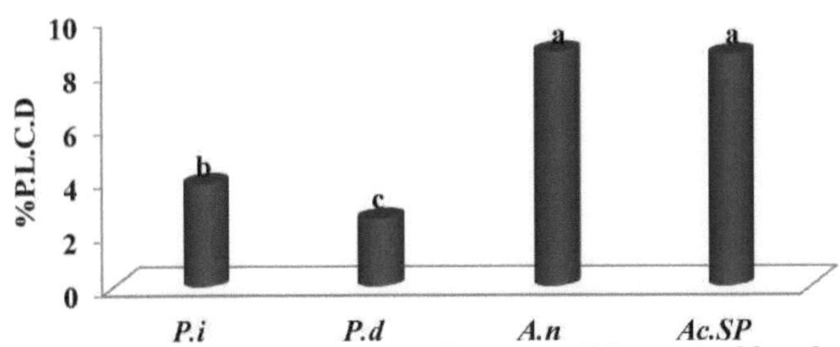

Figure 2: Influence of PDA culture media on mycelial growth of four fungal species.

P. i= Pinicillium italicum
A. n= Aspergillus niger
P. d=Pinicilium digitatum
Ac. SP= Acremoniums sp

Effect of temperature

The results of the temperature effect are shown in (Figure 3 to 8). It is clear from this figure that the molds tested belong to two different physiological groups, depending on their optimum growth temperatures. In fact, each fungal species has a temperature range that is suitable for its development; according to the figure below, we distinguish between thermophilic fungi that prefer to grow at high temperatures, such as *Aspergillus niger*, whose optimum temperature is around 30°C, and mesophilic fungi, whose optimum temperatures are between 20 and 25°C, such as *P. italicum* and *P. digitatum*, which thrive best at 20°C, while *A.niger* and *Acremonium sp.* prefer temperatures around 25°C.

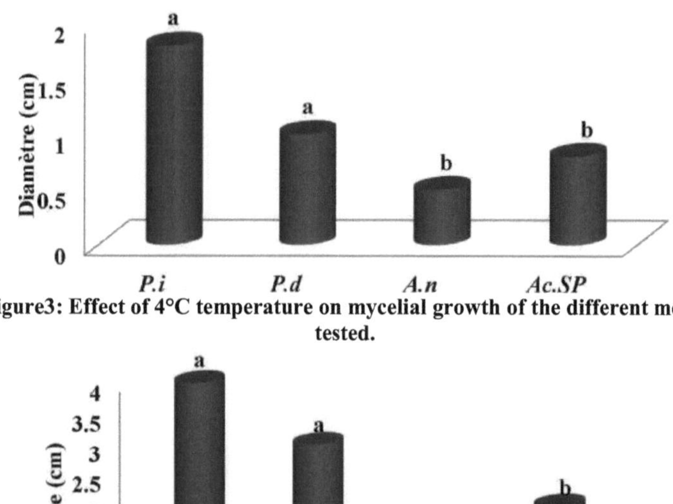

Figure3: Effect of 4°C temperature on mycelial growth of the different molds tested.

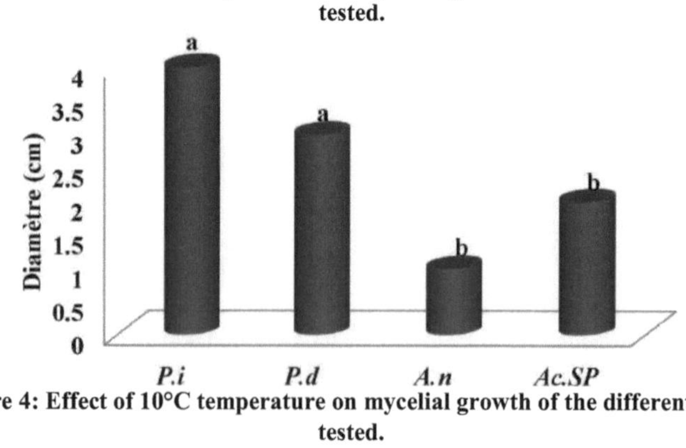

Figure 4: Effect of 10°C temperature on mycelial growth of the different molds tested.

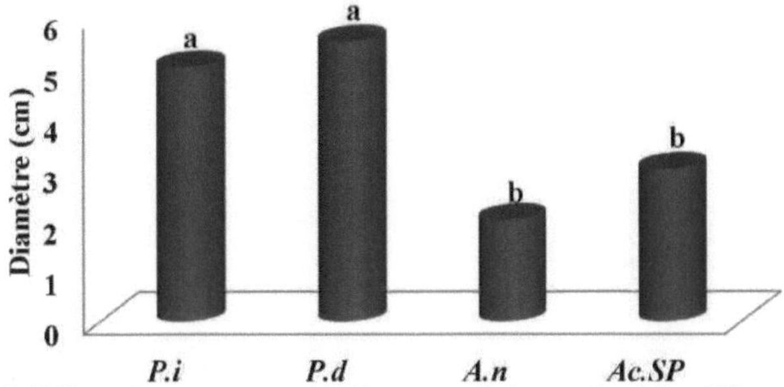

Figure 5: Effect of temperature 15°C on mycelial growth of the different molds tested.

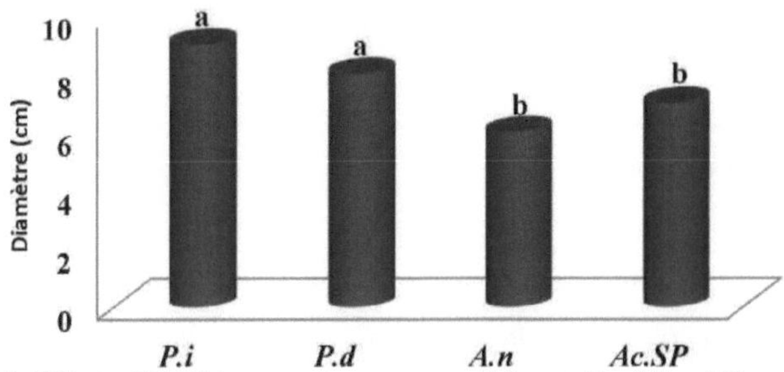

Figure 6: Effect of 20°C temperature on mycelial growth of the different molds tested.

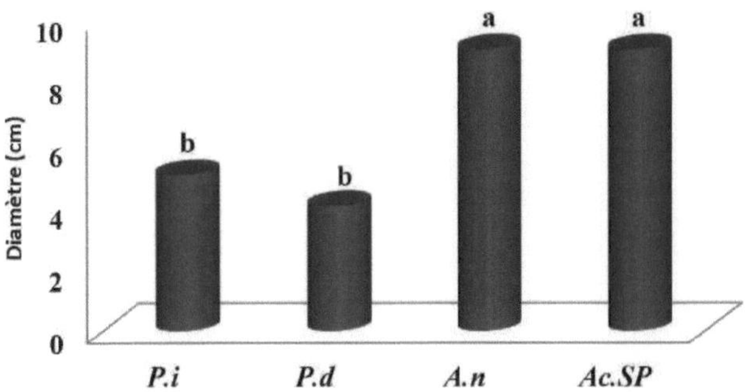

Figure 7: Effect of 25°C temperature on mycelial growth of the different molds tested.

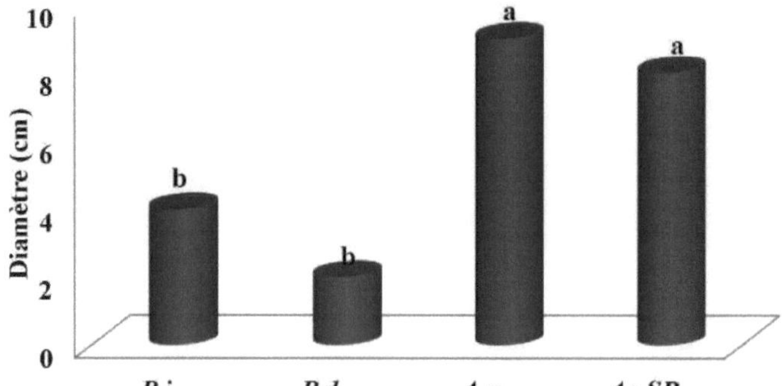

Figure 8: Effect of temperature 30°C on mycelial growth of the different molds tested.

P. i= Pinicillium italicum
A. n= Aspergillus niger
P. d=Pinicilium digitatum
Ac. SP= Acremoniums SP

Several authors have reported that *Penicillium* grow at moderate temperatures of around 20 to 27°C (Moreau, 1980; Molimard et *al.,* 1995; Bensoussan, 2007; Tabuc, 2007); however, the optimum growth temperature for *Aspergillus niger* generally varies between 25 and 30°C; this mold can continue to grow up to 42°C (Tabuc, 2007). As for *Acremoniums,* their optimum growth temperatures range from 20 to 30°C (Peberdy, 2013).

CONCLUSION

The aim of the present work is to find new natural products with antifungal properties against post-harvest citrus deterioration fungi, and to compare the efficacy of these products with that of chemical fungicides used in this field.

Citrus fruits are rich in water and sugars that are easily metabolized by molds. In the first part of this work, we isolated some of the moulds responsible for citrus deterioration. The results of their identification enabled us to recognize the following species: *Penicillium italicum, Penicillium digitatum, Aspergillus niger* and *Acremonium sp.*, all of which were found to cause citrus rot.

To develop properly, fungi require incubation temperatures and culture media of the right chemical composition. These two factors influence the stages of their development cycle. The fungi studied grow well on PDA and MEA media, with slight differences in their growth rates in the two media; similarly, temperature influences the development of *Penicillium italicum, Penicillium digitatum, Aspergillus niger* and *Acremonium sp.* Grow well at 20-25°C, while *Aspergillus niger* growth is maximal at 30°C.

Similar results have been reported for *Alternaria alternata* and *Penicillium expansum* (Selmaoui, 1999). Other studies have shown that optimum germination, mycelial growth and sporulation occur between 25 and 30°C for *Alternaria alternata* (Hajisà, 1970; Lamarque, 1987).

BIBLIOGRAPHICAL REFERENCES

Bensoussan M., Charpentier C., Roux B., Guilmart A. and Dantigny P., 2007; *Global warming, which potential impacts on the vineyards?*

Botton B., Burton A., Fevre M., Gauthier S., Guy P.R., Larpent J.P., Reymond P., Sanglier J.J., Vayssier Y. et Veau P., 1990; Moisissures utiles et nuisibles, importance industrielle. Édition Masson, 511 p.

Chapot, H. 1964. LE citronde borneo. *Al Awamia* 13: 137-153.

Davet P. and Rouxel F., 1997; Détection et isolement des champignons du sol. Édition INRA, Paris, 203 p.

Dilami A., 2016; Special SIAM 2016 ; L'ECONOMISTE.

FAO, 2003; Groupe Intergouvememental sur les Agrumes, Systèmes de Production D'Plants D'Agrumes Sains à Cuba ; ccp: ci 03/10.14p.

Hasijà (S.K.) - Physiological studies of *Alternaria citri* and *A. tenuis*. -*Mycologia*, 1970, **62**(2), 289-295.

Lamarque (C.) - Étude comparée de l'évolution des *Alternaria* du tournesol en Australie et en France. - *Phytoma*, 1987, (393), 24-27.

Ministère de l'Agriculture et de la Pêche Maritime, 2012; Situation de l'Agriculture MAROCAINE - N°10 ; Dossier, Produits de terroir : une stratégie d'action le long de la chaine de valeurs.

Ministère de l'Agriculture et de la Pêche Maritime, 2013; Note de veille secteur agrumicole Novembre 2013; Direction de la Stratégie et des Statistiques ; Note stratégique n°97.

Ministère de l'Agriculture et de la Pêche Maritime, 2014; L'Année Agricole Ju illet 2014; Direction de la Stratégie et des Statistiques ; Note stratégique n°98.

Molimard P., Vassal L., Spinnler H., 1995; Suivi de croissance de *Pénicillium camemberti* et *Geotrichum candidum* en culture pure et en association au cours de l'affinage de fromages expérimentaux à pâte molle de type camembert ; Le Lait, INRA Editions, 1995, 75 (1), pp.316. <hal-00929415>.

Moreau C., 1980; Le Penicillium roqueforti, morphologie, physiologie, intérêt en industrie fromagère, mycotoxines (Révision bibliographique); Le Lait, INRA Editions, 1980, 60 (595 596), pp.254-271. <hal-00928855>.

Peberdy J.F., 2013; *Penicillium and Acremonium. Springer Science & Business Media. Biotechnology handbooks: p. 57. 298pages*

Pitt J.I., 1988; Laboratory guide to common *Penicillium* species. Commonwealth Scientific and Industrial Research Organisation. Division of Food Processing. 187 p.

Selmaoui (K.) - Étude d'un complexe fongique responsable de la pourriture des pommes en conservation. Application de quelques moyens de lutte chimique. *Thesis Doct. Univ. Ibn Tofaïl, Fac Sci. Kenitra, Morocco,* 1999, 175 p.

Samson R.A., Hoekstra E.S. and Frisvad J.C., 2004; Introduction to Food- and Airborne Fungi. Sixth edition. CBS, Utrecht, The Netherlands.

Tabuc C., 2007; Flore Fongique de Différents Substrats et Conditions Optimales de Production des Mycotoxines ; thèse ; Université de Bucarest, Institut National Polytechnique de Toulouse. 190p.

ANTIFUNGAL ACTIVITY OF TWO ESSENTIAL OILS AGAINST CITRUS FRUIT ROT IN MOROCCO

ANTIFUNGAL ACTIVITY OF TWO ESSENTIAL OILS AGAINST THE ROT OF CITRUS FRUIT IN MOROCCO

Summary

Citrus fruit rot is a serious problem, and its frequency increases dramatically from one year to the next, causing considerable damage to orchards and storage facilities, requiring intensive use of antifungal products to combat the pathogens.

The search for new plant protection products of natural origin that are less harmful to human health and the environment has become a necessity. Essential oils (EO), produced by certain plants with antimicrobial properties, can contribute to biological control of citrus spoilage.

The identification results identified two species of *Penicillium, Acremonium* SP and *Aspergillus niger.* The antifungal power of the two essential oils tested *in vitro was* broad-spectrum, with very strong antifungal activity on the diametral growth of pathogens responsible for citrus fruit rot.

Key words; Citrus, Decay, Essential oils, Antifungal activity.

INTRODUCTION

Morocco's citrus sector plays a key socio-economic role, as it is the country's leading fruit export, and is considered a strategic sector contributing to the balance of trade. In recent years, however, Morocco has faced difficulties in maintaining its position on increasingly competitive international markets (El Hadad, 1995).

Citrus is attacked by a number of pathogens from flowering to the harvest stage and subsequently by post-harvest pathogens, most notably fungi, which affect crop production and significantly deteriorate fruit quality (Embaby et *al.,* 2013).

Several types of injury can affect post-harvest fruit quality. Damaged tissue becomes susceptible to infection by pathogenic microorganisms (Fischer et *al.,* 2009). Infection and contamination occur at various stages in the field and after harvest during marketing. Most post-harvest pathogens *(Penicillium, Alternaria, Diplodia* and *Phomopsis)* are able to invade wounds when host defenses are weak. Microorganisms are highly selective in host tissues, depending on pH values (acidic, basic or neutral). Citrus fruits have a pH below 4, so most fungi attack these fruits (Olsen et *al.,* 2000; Milind, 2008).

There are several so-called typical post-harvest diseases, such as green mold, blue mold, *Alternaria* black rot and stem rot, which also reduce fruit quality (Fischer et *al.,* 2009; Embaby et *al.,* 2013).

Green mold (*P. digitatum)* and blue mold (*P. italicum)* are important post-harvest pathogens in citrus. However, both green and blue molds occur in all citrus growing zones and are often the predominant type of rot. (Snowdon, 1990; Naqvi, 2004). These fungi are long-lived and fairly durable, even under adverse conditions. *Penicillium* growth typically occurs following wound infections in produce (Nunes et *al.,* 2010). *Penicillium,* in general, are post-harvest pathogens and a major problem for Mediterranean orchards and the fruit industry (Munoz et *al.,* 2011). Green and blue moulds, respectively, are the main cause of post-harvest losses in oranges. Under Mediterranean climatic conditions, the two together are responsible for 80% of total post-harvest citrus decline (Nunes et *al.,* 2010; Munoz *et al.,* 2011).

Citrus black rot, caused by *Alternaria,* can occur in the field before harvest. The disease is a common problem in Mediterranean climates due to cool, wet winters and hot, dry summers (Brown and Eckert, 2000). In one of the earliest descriptions of *Alternaria* rot of citrus, the causal agent was identified as *Alternaria citri* (Embaby et *al.,* 2013). *Alternaria* rot develops most rapidly in cold storage, and the fungus grows internally along the central stone of the fruit. *Alternaria* rot has been considered a serious problem in lemons and oranges for long-term storage (Embaby et *al.,* 2013).

Citrus rots caused by *Phomopsis citri* are the main postharvest diseases in the citrus-growing area that receive significant rainfall during the fruit development period (Embaby et *al.,* 2013).

During storage periods, these citrus fruits are attacked by a number of moulds, which develop on the fruit, leading to serious economic losses. To minimize the damage caused during the storage phase, the main method used is chemical control, while the industry is increasingly interested in replacing synthetic chemicals with natural products with bioactive properties of plant origin, such as essential oils.

We have described some of the molds isolated from rotting citrus fruit. We then examined the antifungal potential of essential oils, taking as examples the oils of thyme *(Thymus sp.)* and spearmint *(Mentha pulegium).*

MATERIALS AND METHODS

Essential oil extraction processes

There are several methods for extracting EO, each with its own principle, while hydrodistillation and steam distillation remain the most widely used techniques on an industrial scale (Wang et *al.,* 2008). Other techniques include cold expression, volatile organic solvent extraction, superheated steam extraction and supercritical CO_2 extraction (Bruneton, 1999).

Hydrodistillation is the oldest and most cost-effective technique for extracting EOs. Several studies have relied on this technique to extract EOs from thyme (Akrout et *al.,* 2004; El Ajjouri et *al.,* 2008; El-Akhal et *al.,* 2014) and spearmint (Lahrech, 2010; Hmiri et *al.,* 2011; Foganholi et *al.,* 2015); this method has the advantage of not requiring much equipment, and is not expensive (El Ajjouri, 2013).

The extraction of thyme EOs has been the subject of several experimental studies, one of which found that extraction using supercritical CO_2 at low temperature and high pressure gave good results compared with steam stripping (Maldao-Martin et *al.,* 2000).

Culture media

Two culture media were used to grow molds isolated from citrus fruit: PDA and MEA.

PDA (Potato-Dextrose-Agar) medium consists of 200 g potato, 15 g agar and 20 g sucrose, and is mixed in an Erlenmeyer flask with 1000 ml distilled water.

The MEA medium consists of 15 g malt extract, 15 g agar and 1000 ml distilled water (Pitt, 1988; Botton *et al.,* 1990; Davet and Roxel, 1997).

After sterilization in an autoclave at 120°C for 20 min, both media are poured into Petri dishes and inoculated with fungal thallus fragments after solidification, followed by incubation in an oven at 25°C until fungal colonies appear.

If necessary, culture media are supplemented with 100 mg/l chloramphenicol as an antibiotic to inhibit bacterial proliferation.

Isolation and identification of fungal species

Rotten oranges from various cold stores in Kenitra (Figure 1) were disinfected with 75° alcohol. Fragments showing lesions were then cut out and placed in the center of Petri dishes containing PDA medium supplemented with chloramphenicol (Figure 2).

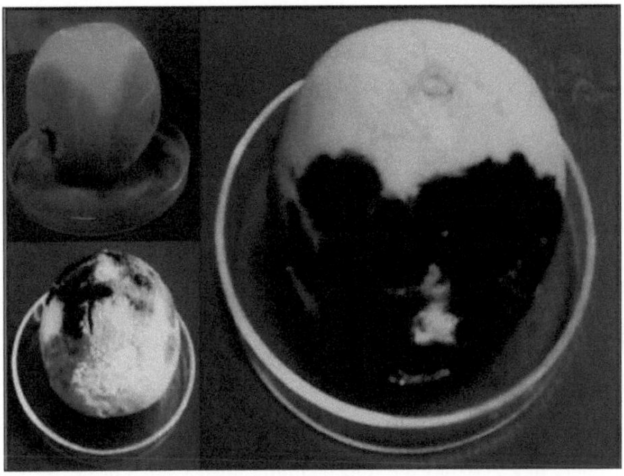

Figure 1: Orange fruit rotted by a fungal complex

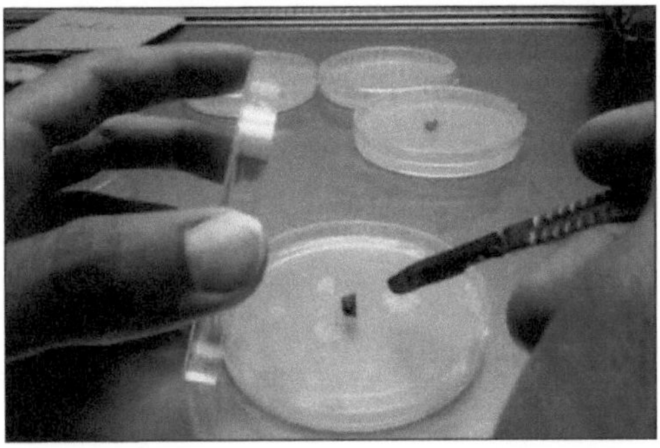

Figure 2: A lesion fragment placed in the center of a Petri dish containing culture medium.

Once the fungal colonies are well developed, a fragment of each colony is subcultured several times in new PDA dishes to obtain pure cultures. Microscope slides are prepared from the pure cultures and observed under a light microscope. The slides are mounted in a drop of distilled water. Isolated fungal species are identified on the basis of their morphological and microscopic characteristics (Samson et *al.,* 2004). Images were taken using an optical microscope (Leica DM750 with built-in camera).

Effect of essential oils on the diametral growth of molds.

The effect of essential oils on mold growth is determined using micro-atmosphere techniques.

A) The micro-atmosphere technique

A.1) Principle of the technique described by Neri* et *al. (2006)

The method chosen to assess the antifungal activity of HE is the micro-atmosphere technique described by Neri et *al.* (2006). In this technique, a Petri dish containing PDA culture medium is first inoculated with a 5 mm diameter fragment taken from the periphery of a 7-10 day old fungal culture; then a small piece of filter paper first soaked with a drop of sterile distilled water is attached to the center of the inside face of the Petri dish lid; the filter paper is impregnated with a certain volume of pure EO (15, 30 and 150 µl). The prepared cultures are then sealed tightly with Parafilm to prevent external contamination and also to limit the escape of EO vapors to the outside of the culture. The culture is then incubated in the dark, lid down, at 25°C. Each trial is repeated three times. A control is prepared under the same conditions, except that the HE is replaced by sterile distilled water (Figure 3).

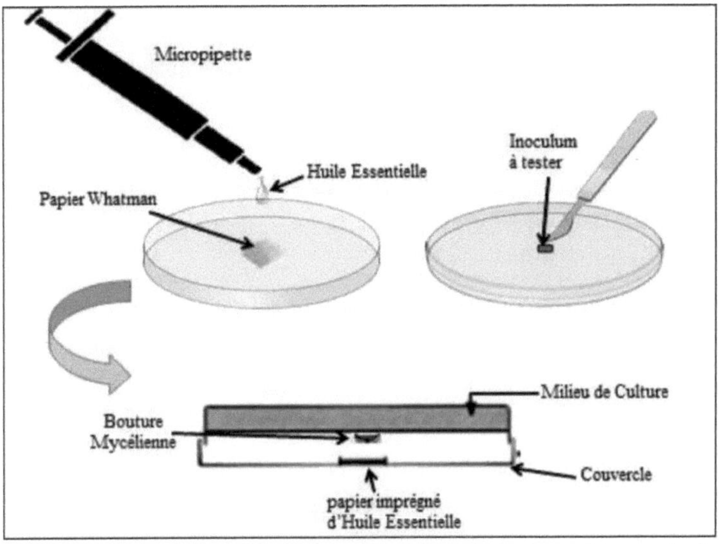

Figure 3: Preparation of the HE test using the micro-atmosphere method

A.2) Reading and expressing results

The rate of fungal diametral growth is determined by repetitive measurements of thallus diameters every 24 h. The effect of HE on mycelial growth is measured by the percentage inhibition of diametral growth (P.I.C.D.), determined by the following relationship:

P.I.C.D = (0t-0e/0t) x 100

With :

0t = Average diameter of control thalli.

0e = Average diameter of thalli confronted with a certain concentration of an essential oil.

B) Dilution preparation

Given the non-miscibility of EO with water and therefore with the culture medium, a homogeneous distribution is achieved using a 0.2% agar solution (SA) to emulsify these compounds. This has the advantage of incorporating these EOs into the culture medium without the need for a foreign diluent; furthermore, the agar chosen is commonly used in microbiology and has no known side effects.

C) Preparation of the stock solution

The stock solution (SM) is 1:25ème , corresponding to the first concentration in the dilution range. It is prepared as follows: a volume of 2.4 ml of the EO to be tested is added aseptically to 57.3 ml of 0.2% agar water (SA) previously autoclaved (120°C) for 20 min. The resulting solution is vigorously vortexed for a few minutes to disperse the oil.

D) Preparation of successive dilutions of the stock solution

Successive dilutions are also made in 0.2% agar solutions. From the stock solution, successive dilutions were made to obtain dilutions of 1:10ème , 1:50ème and 1:100ème . Their concentrations are prepared as follows:

- 1/10 solution: 7 ml essential oil + 63 ml SA,
- 1:50 solution: 50 ml of 1:25 stock solution + 50 ml SA,
- 1/100 solution: 32 ml of 1/50 dilution + 32 ml SA.

- Preparation of culture media containing different concentrations of HE

Under a laminar flow hood, 1.5 ml of the various dilutions prepared (1/10, 1/50, 1/100) are aseptically added to test tubes each containing 13.5 ml of PDA culture medium at 39 g/l, sterilized and kept under supercooling at 45°C in a water bath, to obtain final HE dilutions of 1/100, 1/500 and 1/1000.

Using a vortex, the tubes containing the culture medium with added EO are shaken before being poured into Petri dishes, at a rate of three repetitions per fungus and per concentration.

Controls were also prepared, containing 1.5 ml of 0.2% agar water (SA) and 13.5 ml of culture medium instead of the HE dilutions.

E) Seeding and incubation of cultures

Using a sterile scalpel, fungal culture fragments of around 0.5 cm are harvested from 7 to 10-day-old, fully-grown cultures and aseptically deposited in the center of Petri dishes prepared previously with the desired dilutions.

The seeded Petri dishes are then hermetically sealed with Parafilm, before being incubated in an oven at 25°C in the dark.

F) Reading and expressing results

Results are read and expressed in the same way as for the micro-atmosphere technique.

STATISTICAL ANALYSIS

Statistical processing of the data was carried out using XLSTAT software, with the "Duncan" test at the 5% threshold.

RESULTS AND DISCUSSION

Isolation and identification

Using oranges that had rotted during storage, we were able to isolate and identify several fungal agents responsible for citrus rot. Ten fungal species were identified from the lesions of four rotten oranges: *Penicillium italicum, Penicillium digitatum, Aspergillus niger Acremonium sp.* The identity of all the fungal species isolated will be confirmed by phyllogenetic studies.

Assessment of the sensitivity of each species to the essential oils tested

The antifungal activity of the essential oils of thyme and spearmint is determined by two techniques: the patmosphere technique and direct contact. The volumes tested in the case of the micro-atmosphere technique are: 15µl, 30µl and 150 µl.

For each chart :

- ✓ On the ordinate: % inhibition,
- ✓ Abscissa: incubation time.

a-Penicillium italicum

The results of the in vitro comparison of EO with P. italicum (Figures 4) show that both products exerted satisfactory inhibition, with a slight difference in the degree of efficacy.

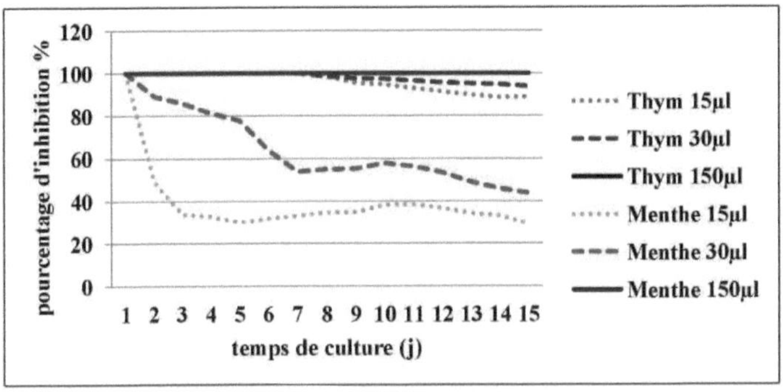

Figure 4: Antifungal activity of thyme and spearmint EOs on the diametral growth of *P. italicum* using the micro-atmosphere technique.

For the micro-atmosphere technique (Figure 4): both EOs gave remarkable inhibition results, the 3 concentrations of thyme exerted a quasi-total inhibition during the first 8 days and which started to decrease up to 90% in the 15ème day; while for mint, total inhibition was obtained only for the 150µl volume, the other two volumes (15 and 30µl) marked a weak inhibition and which decreased as a function of the incubation time to finish around 30% for 15µl and 43% for 30µl.

b-Penicillium digitatum

P. digitatum proved more sensitive to EO, while atmospheric confrontation is still the most effective test.

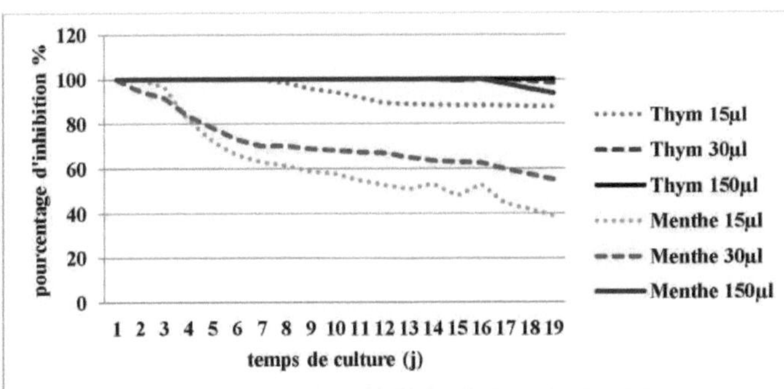

Figure 5: Antifungal activity of thyme and spearmint EOs on the diametral growth of *P. digitatum* using the micro-atmosphere technique.

Similarly for *P. digitatum, the* activity of the two EOs, tested using the microatmosphere technique, remains more effective.

The percentage of inhibition obtained in µatmosphere is 100% for the 3 volumes of thyme tested as well as for the 150µl volume of mint; however in the case of the other two volumes of mint (15 and 30µl), the percentage of inhibition was 100% at the beginning but began to decrease slightly over time to reach 56% and 40% respectively at day 19ème (Figure 5).

c-Aspergillus niger

Aspergillus niger was found to be sensitive to mint and thyme EOs in microatmospheres.

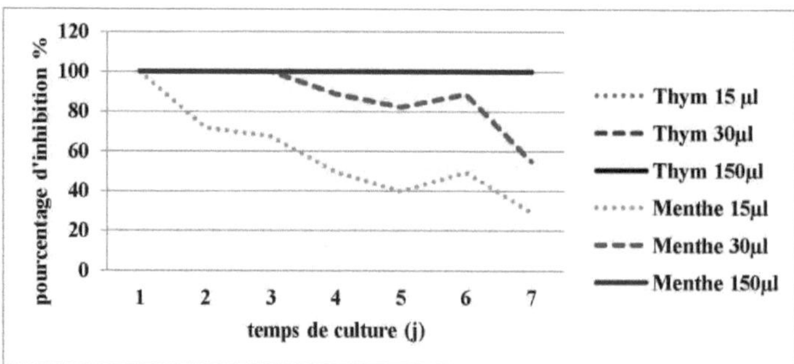

Figure 6: Antifungal activity, using the micro-atmosphere technique, of thyme and spearmint EOs on the diametral growth of *Aspergillus niger.*

The results obtained with the volatile substances of the two EOs are satisfactory, while the inhibition obtained is total for all volumes, this inhibition remains permanent for the 3 volumes of thyme as well as with the 150µl volume of mint, while it drops on day 1 for 15µl and on day 3ème for 30µl of mint to reach respectively 30% and 57% at day 7ème (Figure 6).

d-Acremonium sp.

These graphs show that the *Acremonium* sp. strain is more sensitive to mint and thyme essential oils, and precisely to the micro-atmosphere technique.

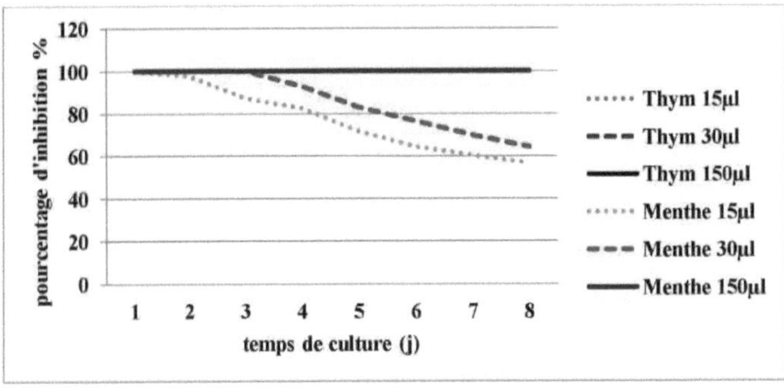

Figure 7: Antifungal activity, using the micro-atmosphere technique, of thyme and spearmint EOs on the diametral growth *of Acremonium* sp.

According to (figure 7), in micro-atmosphere, both EOs showed total inhibition with all volumes, this inhibition remaining permanent for the 3 volumes of thyme as well as for 150µl of mint, while it declines at 2ème day and 3ème respectively for the 15µl and 30µl volumes of mint, ending up at almost 60% for both volumes at 8ème day.

These results suggest the antifungal power of the EOs studied, and their efficacy against our fungal germs, both in their totality and with their volatile substances, while this efficacy is influenced by incubation time, with a negative correlation between the percentage of inhibition and incubation time.

There are two possible explanations for this situation, one being that these declines are caused either by evaporation of the EOs and their escape from the Petri dishes. Or, that these molds are able to develop resistance to these EOs.

CONCLUSION

The aim of the present work is to find new natural products with antifungal properties against post-harvest citrus deterioration fungi, and to compare the efficacy of these products with that of chemical fungicides used in this field.

Citrus fruits are rich in water and sugars that are easily metabolized by molds. In the first part of our work, we isolated some of the moulds responsible for citrus deterioration; the results of their identification enabled us to recognize the following species: *Penicillium italicum, Penicillium digitatum, Acremonium* SP and *Aspergillus niger,* all of which were found to cause citrus fruit rot.

To develop properly, fungi require incubation temperatures and culture media of the right chemical composition. These two factors influence the stages of their development cycle. The fungi studied grow well on PDA and MEA media, with slight differences in their growth rates in the two media.

Study of the antifungal activity of essential oils of spearmint and thyme against isolated molds. The results obtained show that EO and give satisfactory results on the diametral growth of these fungal species tested molds, with their volatile fractions.

These results show that EOs can play an essential role in preserving citrus fruits from pathogenic micro-organisms that can affect the quality of these products.

BIBLIOGRAPHICAL REFERENCES

Akrout, A., Chemli R., Chreif I., et Hammami M., 2004 ; Etude des huiles essentielles de quelques plantes pastorales de la région de Matmata (Tunisie). *Cahiers Options Méditerranéennes* 62: 289-292.

Brown, G.E. and Eckert J.W., 2000; *Alternaria* rot. In: Compendium of Citrus Diseases, 2nded (Timmer, L.W.;Garnsey, S.M. and Graham, J.H. ed) APS Press, St. Paul, MN, USA, pp: 38.

El-Akhal F., Greche H., Ouazzani Chahdi F., Guemmouh R., and El Ouali Lalami A., 2014; Composition chimique et activité larvicide sur Culex pipiens d'huile essentielle de Thymus vulgaris cultivées au Maroc Chemical composition and larvicidal activity of Culex pipiens essential oil of Thymus vulgaris grown in Morocco. Accessed June 23. http://www.jmaterenvironsci.com/Document/vol6/vol6 N1/26-JMES-1061-2014-Alakhal.pdf.

El Ajjouri M., 2013; étude de la composition chimique et de l'efficacité des huiles essentielles de quelques espèces du genre *Thymus* dans la préservation du bois contre les champignons lignivores ; Thèse ; Université Mohamed V, Faculté des Sciences Rabat. 181p.

El Ajjouri M., Satrani B., Ghanmi M., Aafi A., Farah A., Rahouti M., Amarti F., and Aberchane M.. 2008; Antifungal activity of essential oils of Thymus bleicherianus Pomel and Thymus capitatus (L.) Hoffm. & Link against wood-rotting fungi. *Biotechnologie, Agronomie, Société et Environnement* 12 (4): 345-351.

El Hadad F., 1995; Enjeux et perspectives de la filière agrumes du Maroc ; CIHEAM option méditerranéennes. Sér. B/n°14, 250-264p.

Embaby E.M., Hazaa M., Hagag L. F., Ibrahim T. E. and Abd El-Azem F. S., 2013; Decay of Some Citrus Fruit Quality Caused by Fungi; Journal of Applied Sciences Research, 9(11): 5920-5929.

Fischer, I. H., Ferreira M. D., Spósito M. B., and Amorim L., 2009; Citrus postharvest diseases and injuries related to impact on packing lines. *Scientia Agricola* 66 (2): 210-217.

Foganholi Ana Paula Do A. M., de Souza Daniel J.F., Santiago D.C., Orives J. R., Pereira J.P., and De Jesus Faria T., 2015; Composiào quimica e atividade antifùngica do oleo essencial de poejo em diferentes estàgios de desenvolvimento. *Semina: Ci?ncias Agr?rias* 36 (5): 3091. doi:10.5433/1679-0359.2015v36n5p3091.

Hmiri S., Rahouti M., Habib Z., Satrani B, Ghanmi M. and El Ajjouri M., 2011; Évaluation du Potentiel Antifongique des Huiles essentielles de *Mentha Pulegium* et *D'Eucalyptus Camaldulensis* dans la Lutte Biologique Contre les Champignons Responsables de la Détérioration des Pommes en Conservation ; Bulletin de la Société Royale des Sciences de Liège, Vol. 80, 2011, p.824 - 836.

Lahrech K., 2010; Extraction et analyses des huiles essentielles de *menthe pulegium L.* et de *Saccocalyx satureioide* ; Tests d'Activités antibactériennes et antifongiques ; mémoire de magister; Université d'Oran Es-Sénia, Faculté des sciences; 90p.

Maldao-martin M., Palovra A., Beiraro De Costa M.L., Bernardo M.G., 2000; Supercritical Co2 extraction of *Thymus zigis L. Subsp. Sylvestris* aroma. Journal of supercritical Fluids. 18, 25-34. 2000.

Milind S.L., 2008; Citrus fruit biology, technology and evaluation. First ed book, Copyright © Elsevier Inc. All Rights Reserved Academic Press is an Imprint of

Muñoz A., López-García B., Veyrat A., González-Candelas L. and Marcos J.F., 2011; Comparative analysis of the sensitivity to distinct antimicrobials among *Penicillium* spp. causing fruit postharvest decay; *Phytopathol. Mediterr.* (2011) 50, 392-407.

Naqvi, s.a.m.h., 2004; diseases of fruits and vegetables, diagnosis and management (vol i). Kluwer academic publishers, new york, pp: 679.

Neri F., Mari M. & Brigati S., 2006; Control of *Penicillium expansum* by volatile compounds. *Plant Pathol,* 55, 100-105.

Nunes C., Duarte A., Manso T., García J.M., Cayuela J.A., Yousfi K., Martínez, C. Weiland M.C., Salazar M., 2010; Relationship between Postharvest Diseases Resistance and Mineral Composition of Citrus Fruit; Proc. VI IS on Mineral Nutrition of Fruit Crops; Eds.: M. Pestana and P.J. Correia; Acta Hort. 868, ISHS 2010.

Olsen M., Matheron M., Mcclure M., Xiong Z., 2000; Diseases of Citrus in Arizona; The University of Arizona Cooperative Extension.

Samson R.A., Hoekstra E.S. and Frisvad J.C., 2004; Introduction to Food- and Airborne Fungi. Sixth edition. CBS, Utrecht, The Netherlands.

Snowdon, A.L., 1990; Citrus Fruits. Page 70 in: A Color Atlas of Post-Harvest Diseases & Disorders of Fruits & Vegetables. vol. 1: General Introduction & Fruits. CRC Press, Boca Raton, FL.

Wang J., Sun B., Cao Y., Tian Y., Li X., 2008; optimisation of ultrasound-assisted extraction of phenolic compounds from wheat bran- Food Chemistry: Vol. 106: Pp 80-810. 2008.

MIX
Papier aus verantwortungsvollen Quellen
Paper from responsible sources
FSC® C105338

Printed by Books on Demand GmbH, Norderstedt / Germany